MW01595813

For information regarding permission:
AppleTree Institute
415 Michigan Ave., NE
Washington, DC 20017

Printed and bound in Canada.
ISBN 978-1-940641-03-4

AppleTree

Cathryn O'Sullivan is an early childhood educator from Jamaica. She has worked as a teacher, instructional coach, and curriculum writer. In addition to the five books she has published with AppleTree, including *The Doctor* and *James Tries His Best*, she has also published children's songs and books in the Caribbean. Her favorite book as a child was *The Biggest, Most Beautiful Christmas Tree* by Amye Rosenberg.

Christine Llorente is a graphic designer and illustrator located in Miami, Florida. She earned her Bachelor of Fine Arts degree in graphic design and illustration along with a minor in art history from the University of Miami. Christine has worked for Houghton Mifflin Harcourt Publishing, where she designed various programs in both reading and math for grades K-6. She currently runs her own studio, Llorente Design, where she specializes in branding and packaging, pattern and textile design, invitations, photo booth props, and anything else made of paper. Her favorite book as a child was *Anne of Green Gables*.

The Doctor

Written by Cathryn O'Sullivan

Illustrated by Christine Llorente

I don't feel well today.
I can't go outside to play.
I don't want to tell Daddy,
because I know he'll say,

"You need to visit the doctor."

3

When I sneeze, Daddy starts to get a clue.

He says, "Your nose is as red as a parrot at the zoo."

He thinks I might have the flu.

It's time to visit the doctor.

What if the doctor gives me a shot?

What if he looks at my body and finds a great big spot?

What if the doctor can't figure out what I've got?

I don't want to visit the doctor.

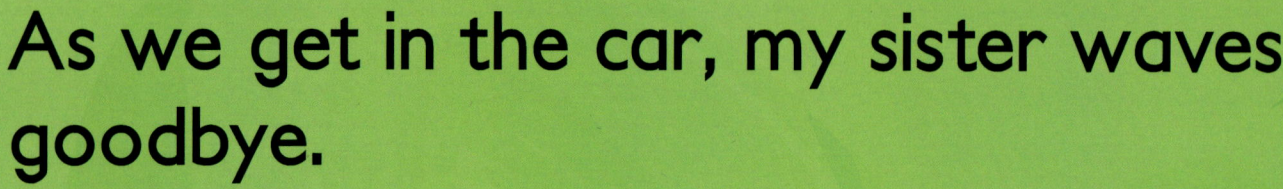

As we get in the car, my sister waves goodbye.

I crawl into my seat with a great big sigh.

All I want to do is cry and cry.

I'm afraid to see the doctor.

Sitting in the waiting room, I know what's in store:

stethoscope, needles, and medicines galore.

Then, the dreaded doctor opens the door...

It's too late now. I'm going to see the doctor.

The doctor lets me in with a big smile.
Maybe it really won't be so vile.
I'll let you know in a little while!

I think I might like the doctor.

The doctor tells me to open up wide,
but I'm not sure. I think I'd rather hide.
He says my mouth and throat look good
inside.

I'm relieved to hear that from the doctor.

Then, the doctor gently looks in my ears
and says, "There's nothing to fear.
You simply have a cold, my dear!"

I even get a treat from the doctor.

I arrive home, and my sister looks at me with pride.

She says I'm the bravest person she's ever eyed.

She says, "I would have wanted to run and hide."

She congratulates me for going to
the doctor.

I listen to what the doctor said.
I take it easy and stay in bed.
Think of all the playing that's now ahead!

Everyone needs to visit the doctor.